DRONES ARE EVERYWHERE!

DRONES FOR THE POLICE AND MILITARY

DANIEL R. FAUST

PowerKiDS press.
New York

Published in 2020 by The Rosen Publishing Group, Inc.
29 East 21st Street, New York, NY 10010

First Edition

Editor: Shannon Harts
Book Design: Tanya Dellaccio

Photo Credits: Cover (background) GregDPhotos/Shutterstock.com; cover (drone) Marco Cantile/LightRocket/Getty Images; series background Djem/Shuttertock.com; p. 5 (main) Adam Melnyk/Shutterstock.com; p. 5 (inset) Parilov/Shutterstock.com; p. 7 https://upload.wikimedia.org/wikipedia/commons/9/9e/9th_Attack_Squadron_-_Trainer.png; p. 9 JACK GUEZ/AFP/Getty Images; p. 11 Eugene Berman/Shutterstock.com; p. 12 VanderWolf/Shutterstock.com; p. 13 Valery Matytsin/TASS/Getty Images; p. 15 Isaac Brekken/Getty Images News/Getty Images; p. 17 CapturePB/Shutterstock.com; p. 19 MediaNews Group/The Mercury News/Getty Images; p. 21 Michael Nagle/Getty Images News/Getty Images; p. 22 picture alliance/Getty Images.

Cataloging-in-Publication Data

Names: Faust, Daniel R.
Title: Drones for the police and military / Daniel R. Faust.
Description: New York : PowerKids Press, 2020. | Series: Drones are everywhere! | Includes glossary and index.
Identifiers: ISBN 9781725309340 (pbk.) | ISBN 9781725309364 (library bound) | ISBN 9781725309357 (6 pack)
Subjects: LCSH: Drone aircraft–Juvenile literature. | Military robots–Juvenile literature. | Law enforcement–Juvenile literature.
Classification: LCC UG1242.D7 F38 2020 | DDC 623.74'69–dc23

Manufactured in the United States of America
CPSIA Compliance Information: Batch #CWPK20. For Further Information contact Rosen Publishing, New York, New York at 1-800-237-9932.

CONTENTS

LOOK UP! 4
HOW DRONES WORK............ 6
TYPES OF DRONES 8
DANGEROUS JOBS 10
EYES IN THE SKY.............. 12
SEARCH AND DESTROY......... 14
MANHUNTERS 16
LOST AND FOUND 18
DRONES ON THE GROUND 20
DRONES OF TOMORROW...... 22
GLOSSARY................... 23
INDEX 24
WEBSITES 24

LOOK UP!

You've probably heard the word "drone" in the news. Either a **pilot** on the ground directs these aircraft using a **remote control**, or an onboard computer controls them.

Although flying drones has become a popular hobby for many, the military has historically used them. Drones are perfect for dangerous, or unsafe, tasks such as flying over enemy territory for **surveillance**, searching for survivors after a natural disaster—a sudden and terrible event in nature—or handling **hazardous** matter.

DRONE DETAILS

A DRONE IS ALSO CALLED AN UNMANNED AERIAL VEHICLE, OR UAV. AERIAL MEANS IN THE AIR, AND A VEHICLE IS A MACHINE THAT TAKES PEOPLE OR THINGS FROM ONE PLACE TO ANOTHER.

Drones are all over the news these days. In December 2018, hundreds of flights were canceled at Gatwick Airport in London, England, because drones were flying over the airfield.

HOW DRONES WORK

Drones can come in many shapes and sizes, but they all have the same basic **components**. Drones are usually made of plastics or lightweight metals, such as aluminum. A battery, or small object that provides a larger object with electricity, often supplies power to the drone.

Most drones have simple **navigational** equipment, or tools. Many also come with cameras and other sensors. Drones are directed using a small, handheld object that looks like a video game controller, easy-to-carry computers, or large command stations.

DRONE DETAILS

THE FIRST DRONES THE MILITARY USED WERE AERIAL TORPEDOES—OR TUBE-SHAPED BOMBS. THEY WERE SELF-MOVING TORPEDOES AND AIRPLANES COULD DROP THEM INTO THE WATER.

Although the military flies drones all over the world, many are controlled from computers at certain military bases.

TYPES OF DRONES

A drone can be anything from a radio-controlled flying toy to a 32,000-pound (14,515 kg) aircraft. Some drones are small enough to fit in your hand. A lot of the drones local law enforcement use are like the kinds of drones you can buy online or at a store.

Military drones are much more **sophisticated**. These drones, like the MQ-1B Predator or MQ-9 Reaper, have advanced sensors and surveillance tools. Many military drones like these are armed.

DRONE DETAILS

BEFORE THEY STARTED FLYING THEIR OWN DRONES, POLICE ACROSS THE COUNTRY WOULD BORROW DRONES FROM GOVERNMENT DEPARTMENTS, LIKE THE FBI AND THE DEPARTMENT OF HOMELAND SECURITY.

Some of the smallest drones the military uses can be carried and launched, or sent off, by hand—like this RQ-11 Raven.

DANGEROUS JOBS

Whether they are flying over enemy territory or searching the **rubble** after a disaster, one of a drone's main roles is to accomplish jobs that may be too dangerous for human pilots. The U.S. military began gathering information on ways to use unmanned aircraft to spy on, or secretly watch, enemies after the **Soviet Union** shot down a U-2 spy plane in 1960.

Military and law enforcement officials around the world continue to find ways to use unmanned vehicles in dangerous settings to save people's lives.

DRONE DETAILS

AN EARLY USE OF AERIAL VEHICLES FOR SURVEILLANCE OCCURRED DURING THE AMERICAN CIVIL WAR. BOTH SIDES USED HOT AIR BALLOONS TO SPY ON EACH OTHER.

Before drones, spy planes like the U-2 were in charge of surveillance. In 1960, a U-2 was shot down over the Soviet Union and the pilot, Francis Gary Powers, was taken prisoner.

EYES IN THE SKY

The main missions, or tasks, of military and police drones have always been **reconnaissance** and surveillance. Drones with cameras can take both video and still pictures. Heat-sensing cameras and radar, a method of finding faraway objects using radio waves, allow them to work at night or during bad weather.

Because drones do not carry human pilots, they can operate for 24 hours a day, seven days a week. A single drone can circle an area continuously, providing up-to-the-minute reports to its operators on the ground.

Military UAVs have very advanced cameras that can capture pictures both day and night.

SEARCH AND DESTROY

Despite what you may have seen in the news, on TV, or in movies, many military drones are not armed. Armed drones are called unmanned combat aerial vehicles, or UCAVs. They are sometimes called hunter-killers.

Combat drones are usually armed with missiles, or objects that are shot or launched to strike from faraway. They also have advanced sensors and computers for targeting, or aiming. Because they are built to carry heavy weapons, combat drones are some of the largest drones in the world.

DRONE DETAILS

WHILE ARMED DRONES HAVE SAVED COUNTLESS AMERICAN LIVES, THEY HAVE ALSO KILLED HUNDREDS AND MAYBE EVEN MORE THAN 1,000 CIVILIANS, OR THOSE NOT IN THE MILITARY, ACCORDING TO FACTS SOME JOURNALISTS HAVE COLLECTED.

The MQ-9 Reaper is a hunter-killer drone. It's built for combat, or active fighting, and includes all the tools needed to locate and destroy its target.

MANHUNTERS

Since the war on terror began in 2001, the U.S. military has used its drone fleet to search for terrorists, people who use violence to scare others to achieve political goals, around the world.

State and local police also use drones with cameras to help them find criminals and **fugitives** for several reasons. Drones can be cheaper than other aircraft. Drones also allow police to search a large area in less time than human officers could search on foot.

DRONE DETAILS

WITH MORE CAMERA-EQUIPPED DRONES IN THE SKY, MANY CITIZENS ARE WORRIED ABOUT THEIR PRIVACY. SINCE 2013, ABOUT 41 STATES IN THE UNITED STATES HAVE CREATED LAWS REGULATING, OR CONTROLLING, DRONE USE.

It can take police officers days or even weeks to search everywhere a criminal might be hiding. Drones help the police search a large area in less time, keep officers safe, and get dangerous people off the streets quicker.

LOST AND FOUND

Drones can be used to find more than just criminals. They can also help find runaways, lost hikers, and other missing people.

A single drone with a camera can search a large area from above in a matter of minutes, while a human search party would take several hours to search the same area. Drones can search day and night, and see through clouds, rain, and fog. Drones can even deliver food, water, and first-aid to people trapped in unsafe and remote, or out-of-the-way, places.

DRONE DETAILS

DRONES HAVE BECOME AN IMPORTANT TOOL FOR HELPING AFTER NATURAL DISASTERS LIKE FLOODS AND TORNADOES, OR STORMS WITH VERY STRONG WINDS THAT MOVE AROUND A CENTRAL POINT. THEY CAN ALSO SEARCH FOR SURVIVORS IN DAMAGED, UNSAFE BUILDINGS.

Police can use drones to fly over an area after a flood to find people who may need to be rescued, or saved.

DRONES ON THE GROUND

Not all drones fly. It's true! Drones that operate on land are called unmanned ground vehicles, or UGVs. Some UGVs are small and look like remote-controlled cars and others are almost as large as tanks, or heavily-armored military vehicles that move on two large belts with wheels inside.

Like their flying cousins, UGVs are used in conditions that might be too dangerous for humans. Some UGVs are used to locate hidden bombs. Other kinds of UGVs are sent ahead of teams of soldiers to check for enemy troops.

DRONE DETAILS

DURING WORLD WAR II, THE GERMAN ARMY USED A REMOTE-CONTROLLED EARLY UGV. THE GOLIATH LOOKED LIKE A SMALL TANK AND COULD CARRY AS MUCH AS 220 LBS (99.8 KG) OF HIGH-EXPLOSIVES.

Police departments use larger UGVs, like the Andros, to safely get rid of bombs.

DRONES OF TOMORROW

It's very unlikely that the military and police will stop using drones any time soon. We are likely going to see more drones built to handle many kinds of missions, or tasks. The U.S. military is experimenting with tiny, insect-sized drones that could be sent out as a swarm, or huge crowd, in enemy territory.

The U.S. Navy is now using underwater drones, or unmanned undersea vehicles (UUVs) that can be sent out from a submarine, or underwater ship, to explore ocean **depths** and find underwater mines.

GLOSSARY

component: A part of a larger whole.

depth: An area that exists far below the surface.

fugitive: A person who has run away to avoid being captured.

hazardous: Involving risk or danger.

navigational: Relating to the act or process of finding a way from one place to another.

pilot: A person who flies an aircraft.

reconnaissance: The exploration of a place to collect information.

remote control: A small machine that's used to control another machine from a distance.

rubble: Broken pieces of stone, brick, or other materials from walls or buildings that have fallen.

sophisticated: Highly developed and complex.

Soviet Union: A former country in eastern Europe and northern Asia.

surveillance: The act of watching someone or something closely.

INDEX

B

battery, 6

C

camera, 6, 12, 13, 16, 18

H

hunter-killer, 14, 15

P

pilot, 4, 10, 11, 12
Powers, Francis Gary, 11

R

reconnaissance, 12
remote control, 4, 20

S

sensors, 6, 8, 14
surveillance, 4, 8, 10, 11, 12

T

torpedoes, 6

U

unmanned aerial vehicle (UAV), 4, 13
unmanned combat aerial vehicle (UCAV), 14
unmanned ground vehicle (UGV), 20, 21
unmanned undersea vehicle (UUV), 22

WEBSITES

Due to the changing nature of Internet links, PowerKids Press has developed an online list of websites related to the subject of this book. This site is updated regularly. Please use this link to access the list: www.powerkidslinks.com/dae/police